AF119253

YOUR KNOWLEDGE HAS VALUE

- We will publish your bachelor's and master's thesis, essays and papers

- Your own eBook and book - sold worldwide in all relevant shops

- Earn money with each sale

Upload your text at www.GRIN.com
and publish for free

Bibliographic information published by the German National Library:

The German National Library lists this publication in the National Bibliography; detailed bibliographic data are available on the Internet at http://dnb.dnb.de .

This book is copyright material and must not be copied, reproduced, transferred, distributed, leased, licensed or publicly performed or used in any way except as specifically permitted in writing by the publishers, as allowed under the terms and conditions under which it was purchased or as strictly permitted by applicable copyright law. Any unauthorized distribution or use of this text may be a direct infringement of the author s and publisher s rights and those responsible may be liable in law accordingly.

Imprint:

Copyright © 2015 GRIN Verlag, Open Publishing GmbH
Print and binding: Books on Demand GmbH, Norderstedt Germany
ISBN: 9783668260122

This book at GRIN:

http://www.grin.com/en/e-book/322063/high-dimensional-spatial-indexing-using-space-filling-curves

Ankush Chauhan, William Johnson, Anjuli Patel

High Dimensional Spatial Indexing using Space-Filling Curves

GRIN Publishing

GRIN - Your knowledge has value

Since its foundation in 1998, GRIN has specialized in publishing academic texts by students, college teachers and other academics as e-book and printed book. The website www.grin.com is an ideal platform for presenting term papers, final papers, scientific essays, dissertations and specialist books.

Visit us on the internet:

http://www.grin.com/

http://www.facebook.com/grincom

http://www.twitter.com/grin_com

High Dimensional Spatial Indexing using Space-Filling Curves

William Johnson[1], Ankush Chauhan[2], and Anjuli Patel[3]

Abstract— Representation of two dimensional objects into one dimensional space is simple and efficient when using a two coordinate system imposed upon a grid. However, when the two dimensions are expanded far beyond visual and sometimes mental understanding, techniques are used to quantify and simplify the representation of such objects. These techniques center around spatial interpretations by means of a space–filling curve. Since the late 1800's, mathematicians and computer scientists have succeeded with algorithms that express high dimensional geometries. However, very few implementations of the algorithms beyond three dimensions for computing these geometries exist. We propose using the basic spatial computations developed by pioneers in the field like G. Peano [1], D. Hilbert [2], E. H. Moore [3], and others in a working model. The algorithms in this paper are fully implemented in high-level programming languages utilizing a relation database management system. We show the execution speeds of the algorithms using a space–filling curve index for searching compared to brute force searching. Finally, we contrast three space–filling curve algorithms: Moore, Hilbert, and Morton [4], in execution time of searching for high dimensional data in point queries and range queries.

I. INTRODUCTION AND BACKGROUND

Space–filling curves (SFC) enable higher dimensional objects to be expressed, stored, analyzed, and categorized in one dimensional space. Utilizing a relational database model, there are numerous algorithms to realize such expressions of these objects. When D. Hilbert in 1891 [2], published his paper on mapping a curve to a surface as interpretation of its shape, as shown in Fig. 2, this opened the way for mathematical understandings of higher dimensions to be realized in concrete algorithms. The gift of mathematical expression and algorithms in computer science are applicable to areas of spatial data representation and transformation. Many application areas exist that employ SFC such as two dimensional image compression [5] where Ouni, et al. attempt to utilize a Peano-Hilbert curve to achieve a lossless image compression. Pamula proposes using an SFC to compute wavelet transformations of images [6] deployed in two dimensional road traffic images. However, Pamula does not implement any code to test the inclusion of the Hilbert curve [6]. Newer applications of SFC include the incorporation of muti-dimensional SFC [7] along with hypercubes and

bit-streaming to create more efficient storage structures [7]. Schrack et al. propose using the Morton SFC or *Z-Order curve* as a basis to generate combinations and permutations of new and untested SFC [8] by using the bit interleaving property that dominates the Morton SFC.

Various implementation strategies have been presented in the recent past for high dimensional indexing such as iDistance [9] and iMinMax [10] consisting of data driven approaches. However, a few strategies for space driven approaches in higher dimensions are found, which promotes data reuse using locality properties of SFC.

H. V. Jagadish in 1990 [11] introduced the mapping of multi-dimensional space to a line, based on the Hilbert Curve demonstrating its context with databases and also performing an exhaustive algebraic analysis and computer simulation to show the Hilbert Curve performed equivalent or better than other mappings such as Column-wise, Column-wise snake, Morton-curve , and Gray Code. A similar trend was presented for the Hilbert Curve by M. F. Mokbel [12], the insights from the experiments can be useful for SFC selection depending upon the application usage, but it lacked a comparative analysis for a variety of other SFC with alterations in dimensions and data set sizes.

This paper explores higher dimension implementations of three different, well known SFC algorithms, namely, Hilbert, Morton, and Moore. We compare the execution times of these three SFC related to point and range queries. We reserve kNN queries and implementation of another SFC for future work.

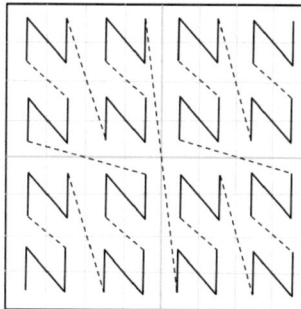

Fig. 1. The third order of a Morton curve.

*This work was not supported by any organization
[1]W. Johnson is a doctoral student with the department of Computer Science, Georgia State University, 25 Park Place South SE, Unit 700, Atlanta, GA, 30303, USA wjohnson6@student.gsu.edu
[2]A. Chauhan is a masters student with the department of Computer Science, Georgia State University, 25 Park Place South SE, Unit 700, Atlanta, GA, 30303, USA achauhan4@student.gsu.edu
[2]A. Patel is a graduating masters student with the department of Computer Science, Georgia State University, 25 Park Place South SE, Unit 700, Atlanta, GA, 30303, USA apatel@student.gsu.edu

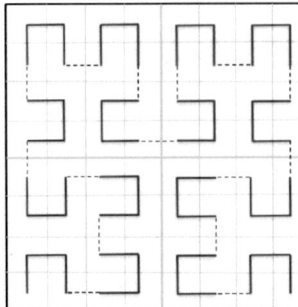

Fig. 2.   The third order of a Hilbert curve.

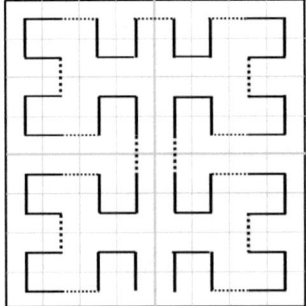

Fig. 3.   The third order of a Moore curve.

II. IMPLEMENTATION

We compare the Hilbert SFC with two other SFC, namely, Morton [4], and Moore [3] in numerous realizations that include metrics on performance in generating the SFC on 1K, 10K, 100K, 1M data of uniform and clustered sets. In addition we generate an SFC index to look–up fifty point queries in two, four, eight, and sixteen dimensions. The execution times for point queries using the SFC look–up, are normalized by average millisecond (μ) time and dividing by the average brute force (μ) time.

$$\mu(avg) = \frac{\mu(SFC)}{\mu(bruteforce)} \qquad (1)$$

The analysis of range queries in two and four dimensions are explored and compared using four minimum bounding regions (MBR). We show algorithm execution times and data query times to conclude their performance contrasted against each other. The motivation for these comparisons are to analyze their higher dimensional integrity, construction, speed, and usage for data retrieval.

A. Data Set

A random generator, written in Java was used to generate 1K, 10K, 100K, 1M uniform and clustered data points in a two, four, eight, and sixteen dimensional space. For 1K and 10K, the range of numbers in any dimension is from [0,255]. However, in order to achieve 100K points in other dimensions, the range of numbers in a single dimension was increased to [0,1023]. This range is applied to 100K and 1M for all dimensions. In order to create cluster data, the generator created four centers and applied a Gaussian distribution around each of the centers with a standard deviation of 25 points for 1K and 10K and 100 for 100K and 1M data points. The standard deviation was selected based on the range in a single dimension so that an even number of points could be generated around a center without repeats. Cluster data was not generated for 1M due to the centers have to be strategically placed in the space in order to create four clusters.

For point query, 50 points from the 10K uniform data set were chosen at random. Cluster data was not selected since the test is to determine the speed of retrieval of each point on the space filling curve. Range queries were selected at random for each of the dimensions. Each range query is run against the 10K uniform data set.

B. Hilbert Code Space–Filling Curve

A. R. Butz [13] and later, J. K. Lawder [14], presented mathematical functions proving the continuous bijection for mapping Hilbert SFC beyond two dimensions. The latter is mostly credited for the fulfillment of the Hilbert mapping algorithm.

If we take N bits of a Grey Code as coordinates in N dimensional space we can project a Hilbert SFC on a "walk" using Gray Code on a unit N cube. In this walk, the first step is along the y axis, but we say the walk as a whole travels along the x axis. Travel here specifies the difference between the start and end of the walk of a cube. When $N \geq 2$, the travel of a unit N cube walk is always along a different axis than its first step. We use recursion for $N \geq 2$ so that a long walk is composed of four small Hilbert walks in each quadrant. This means the walk can be achieved by building parent N cube out of 2^N subcubes. Here, bit precision for each point is not required to be explicitly stated for mapping. Rather the bit precision is calculated within the algorithm for the specified the grid edge length, λ, allows us to find the number of recursive steps for a N dimensional space, thus giving the order ω of the curve.

$$\omega = \log_4 N^\lambda \qquad (2)$$

Therefore, the dense-mapping of grid points the order is calculated first in the procedure *HSFC*, which iterates upon all the input data points, calculates their Hilbert SFC index values using a function *HilbertIndex* with arguments containing dimension of the point, calculated order, and coordinates in each of the dimensions.

Instead of filling a finite region of space with infinite detail, the function takes a Hilbert Walk [14] each consisting of unit steps in the N dimensional grid, gradually covering all the intersection points in the grid to their index values. Coordinates of the points are converted into binary using function *ToBinary*, and the bottom left corner of the parent has *BitRange* (0, 0, 0, ... 0), following which the most-significant bits of the coordinates are walked in a unit cube walk. The N most significant bits of a point's index are calculated by a modified inverse *GrayCode* to change each of its N coordinates to the most-significant bits arranged in the direction of walk. The subcubes control less significant bits of the point coordinates.

Now we loop to calculate the subcube with the same recursive steps and using its orientation as an input for the next subcube and calculating the index value, which is achieved by the helper functions *Transform*, *LeftRotation* and *RightRotation*. After all the recursive steps ω are completed for each subcube, the index is modified and returned back to the parent procedure *HSFC*.

Algorithm 1 Bit Range

1: **function** BITRANGE($x, width, start, end$)
2:     **return** $x \gg (width - end) \& ((2**(end - start)) - 1)$
3: **end function**

Algorithm 2 Gray Code

1: **function** GRAYCODE(x)
2:     **return** $x ** (x \gg 1)$
3: **end function**

Algorithm 3 Inverse Gray Code

1: **function** IGRAYCODE(x)
2:     **if** $x = 0$ **then**
3:         **return** x
4:     **end if**
5:     $m \leftarrow \lceil log_2 x \rceil$
6:     $i \leftarrow x$
7:     $j \leftarrow 1$
8:     **while** $j < m$ **do**
9:         $i \leftarrow i ** (x \gg j)$
10:        $j \leftarrow j + 1$
11:     **end while**
12:     **return** i
13: **end function**

Algorithm 4 Transform

1: **function** TRANSFORM($entry, direction, width, x$)
2:     assert($x < 2**width$)
3:     assert($entry < 2**width$)
4:     **return** $RightRotation((x ** entry), direction + 1, width)$
5: **end function**

Algorithm 5 Left Bit-Rotation

1: **function** LEFTROTATION($x, i, width$)   ▷
    $Width : The bit width of x.$
2:     assert($x < 2**width$)
3:     $i \leftarrow i$ mod $width$
4:     $x \leftarrow (x \ll i) | (x \gg width - i)$
5:     **return** $x \& (2**width - 1)$
6: **end function**

Algorithm 6 Right Bit-Rotation

1: **function** RIGHTROTATION($x, i, width$)   ▷
    $Width : The bit width of x.$
2:     assert($x < 2**width$)
3:     $i \leftarrow i$ mod $width$
4:     $x \leftarrow (x \gg i) | (x \ll width - i)$
5:     **return** $x \& (2**width - 1)$
6: **end function**

Algorithm 7 Entry

1: **function** ENTRY(x)
2:     **if** $x = 0$ **then**
3:         **return** x
4:     **end if**
5:     **return** $GrayCode(2 * ((x - 1)/2))$
6: **end function**

Algorithm 8 Hilbert Space-filling Curve

1: **procedure** HSFC($nDim, iGridEdge, iNumValues$) ▷
    Generates the index for iNumValues points in nDim dimensional space.
2:     $totalPoints \leftarrow nDim ** iGridEdge$ ▷ Calculating the total points.
3:     $order \leftarrow log(totalPoints, 4)$ ▷ Calculating the curve order.
4:     $rows[] \leftarrow DB[nDim, iNumValues]$ ▷ Loading the rows from DB
5:     $ctr \leftarrow 0$
6:     **while** $ctr < iNumValues$ **do**
7:         $ctr \leftarrow ctr + 1$
8:         $row[id, dim_1, dim_2, ..., dim_{nDim}] \leftarrow rows[ctr]$ ▷ Load a single tuple.
9:         $id \leftarrow row[id]$
10:        $point \leftarrow row[dim_1, dim_2, ..., dim_{nDim}]$ ▷ Load the point coordinates.
11:        $dim_{1,2,...,nDim} \leftarrow row[]$
12:        $vKey \leftarrow HilbertIndex(nDim, order, point)$ ▷ Calculate the Index.
13:        $DB[iDim, iNumValues] \leftarrow [id, vkey]$ ▷ Load the index in the DB.
14:     **end while**
15: **end procedure**

Algorithm 9 Point to Hilbert Curve Index

1: **function** HILBERTINDEX($nDim, order, point[]$)
2:     $index \leftarrow 0$
3:     $epsilon \leftarrow 0$
4:     $delta \leftarrow 1$
5:     $ctr \leftarrow 0$
6:     **while** $ctr < order$ **do**
7:         $lambda \leftarrow 0$
8:         $ctr \leftarrow ctr + 1$
9:         $dim \leftarrow 1$
10:        **while** $dim \leq nDim$ **do**
11:           $dim \leftarrow dim + 1$
12:           $beta \leftarrow BitRange(Point[dim_{nDim-dim}], order, dim - 1, dim)$
13:           $lambda \leftarrow (lambda|(beta \ll dim))$
14:        **end while**
15:        $lambda \leftarrow$ transform(epsilon, delta, nDim, lambda)
16:        $omega \leftarrow iGrayCode(lambda)$
17:        $epsilon \leftarrow edge * LeftRotation(Entry(dim), delta + 1, nDim)$
18:        $delta \leftarrow (delta + direction(omega + nDim))$
19:        $index \leftarrow ((index \ll nDim)|omega)$
20:     **end while**
21:     **return** $index$
22: **end function**

Hilbert Curve testing was done by using a 50 data point query in all N dimensions, sub queried against the corresponding the corresponding N dimension uniform 10K data set. The results are averaged for the total execution time of 50 point queries consisting of execution time for calculating index values and query execution time. The average execution times for using the Hilbert SFC index is shown in *Fig.5* for 50 data point queries and *Fig.8* for the two and four dimensional range queries. The implementations of Hilbert Curve Algorithms 1–9 were done on a dedicated server having 2.40GHz, Intel® Xeon® E5 CPU, 4GB RAM, 60GB SSD.PostgreSQL was used for storing point datasets relationally according to their dimensions, and their corresponding indexing tables. The development was done in PL/Python, a server side programming language for PostgreSQL. The import and export of queries was done via disk, and for processing datasets beyond 100K size cursors were used for fetching the data in partitions to memory.

C. Moore Code Space–Filling Curve

A variant of Hilbert Curve [2] was proposed by E. H. Moore in 1900 [3][15], called Hilbert-Moore or more commonly referred to as a Moore Curve. Moore proposed the curve should start and end in the same cell in a defined space. This characteristic allows for a closed space filling curve, unlike Hilbert and Morton Curves. In order to achieve the closed loop feature that is distinguishable to the Moore Curve, the space must be divided into four quadrants of equal size. The quadrants are split in the following manner: Quadrant 0 is where Hilbert SFC index is equal to zero and is

located, and Quadrant 3 is where the max Hilbert SFC index resides [15]. The partition of the quadrants can be seen in *Fig 4*.

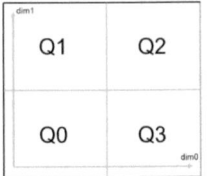

Fig. 4. The quadrant order of a Moore curve.

To compute the SFC index on a Moore Curve, three algorithms (*MooreAP, HilbertCurve, HilbertVkey*) are used to obtain the index. The approach used for creating the Moore Curve begins with generating a Hilbert Curve. A two dimension representation of the Hilbert Curve is determined. Once all points are mapped to a Hilbert index, the algorithm constructs the Moore Curve. Quadrant 0 and Quadrant 3 are flipped (*flip*) along dimension zero, Quadrant 1 is rotated counter-clockwise (*rotateCCW*), and finally Quadrant 2 is rotated clockwise (*rotateCW*) thus creating the *mooreMatrix*. The matrix is then utilized to determine the new Moore index on the data set to create a two dimensional Moore Curve (*MooreAP*). To further extend the dimensions along the Moore Curve, a Hilbert index is calculated with the addition of a new dimension on each iteration until the desired dimension is calculated. Two supplemental functions (*toBinary*, and *toDecimal*) are used to convert from decimal to binary and binary to decimal. This is because there is lack in size constraint on primitive data types in the programming language.

The algorithms were implemented in Java 8 on 2.5GHz Intel i7 CPU, 16GB RAM. Data is stored and analyzed with PostgreSQL 9.5 using a database connection library PostgreSQL JDBC 9.4.1208 that is used to import and export the data from the algorithms. Implementation of higher dimensions of data is a multi–stepped process. Data is first generated, then imported to PostgreSQL database tables and read via the JDBC driver. Next, the algorithms stated above are executed against the data and a Moore index value (*vKey*) is computed for each data point. The nine algorithms below describe the details of the Moore curve generation. Once the *vKey* is generated and stored into a look–up table, then the process of testing begins.

Algorithm 1 Moore Space-filling Curve

1: $iDim \leftarrow I$ ▷ $I=2,4,8,16$
2: $iBitSize \leftarrow K$ ▷ $K=8,10$
3: $table \leftarrow$ DB
4: $sTime \leftarrow$ SystemTime
5: $mooreTable \leftarrow MooreAPCurve(table,mooreMatrix,$ $iDim,iBitSize)$ ▷ mooreMatrix is computation of   ▷ 2D MooreCurve based on bitSize
6: $DB \leftarrow mooreTable$ ▷ mooreTable contains id and vkey on MooreSFC
7: $tTime \leftarrow$ SystemTime - $sTime$
8: **return**

Algorithm 2 MooreAPCurve(*table,mooreMatrix,n,bit*)

1: **for** $i<table.$size() **do**
2:      $mooreTable \leftarrow mooreMatrix$
3: **end for**
4: **if** $n>2$ **then**
5:      **for** $i<n$ **do**
6:          $mooreTable \leftarrow HilbertCurve(mooreTable,n,bit))$
7:      **end for**
8: **end if**
9: **return** $mooreTable$

Algorithm 3 HilbertCurve(*table,n,bit*)

1: **for** $i<table.$size() **do**
2:      $hvkey \leftarrow HilbertVkey(table.$get($i$)$,n,bit)$
3:      $hilbertTable \leftarrow hvkey$
4: **end for**
5: **return** $hilbertTable$

Algorithm 4 HilbertVkey(*point,n,bit*)

$dim0 \leftarrow point.$dim0
$dim1 \leftarrow point.$dim1
$sInterweave \leftarrow dim0,dim1$
$hvkey \leftarrow convertHilbert(sInterweave)$
return $hvkey$

Algorithm 5 rotateCW(*matrix*)

$rotated[][]$
$m \leftarrow matrix.$length()
for $i<m$ **do**
     for $j<m$ **do**
         $rotated[j][m\text{-}1\text{-}i] \leftarrow matrix[i][j]$
     end for
end for
return $rotated$

Algorithm 6 rotateCCW(*matrix*)

1: $rotated[][]$
2: $m \leftarrow matrix.$length()
3: **for** $i<m$ **do**
4:      **for** $j<m$ **do**
5:          $rotated[m\text{-}1\text{-}j][i] \leftarrow matrix[i][j]$
6:      **end for**
7: **end for**
8: **return** $rotated$

Algorithm 7 flip(*matrix*)

1: $flipped[][]$
2: $m \leftarrow matrix.$length()
3: **for** $i<m$ **do**
4:      **for** $j<m$ **do**
5:          $rotated[m\text{-}1\text{-}i][j] \leftarrow matrix[i][j]$
6:      **end for**
7: **end for**
8: **return** $flipped$

Algorithm 8 To_Binary($dim_{1,2,...I},bitSize$)

1: $sBin \leftarrow$ ""
2: **while** $dim_{1,2,...I} >0$ **do**
3:      $sBin \leftarrow (dim_{1,2,...I})$ mod 2
4:      $dim_{1,2,...I} \leftarrow (dim_{1,2,...I}) / 2$
5: **end while**
6: **if** $sBin.$length()$\leq bitSize$ **then**
7:      $leadingZeros \leftarrow bitSize - sBin$
8:      $sBin \leftarrow leadingZeros - sBin$
9: **end if**
10: **return** $sBin$

Algorithm 9 toDecimal(*sBin*)

1: $lDec \leftarrow 0$
2: $p \leftarrow 0$
3: **while** $sBin \neq$ "" **do**
4:      $temp \leftarrow sBin.$lastChar()
5:      $lDec \leftarrow temp * 2^p$
6:      $sBin \leftarrow sBin.$substring($0,sBin.$length()-1)
7:      $p \leftarrow p+1$
8: **end while**
9: **return** $lDec$

Moore Curve testing is done by using a 50 data point query in all N dimensions joined against the corresponding N dimension 10K uniform data set and four MBR queries in two and four dimensions joined to the corresponding N dimension 10K uniform data set. Results are averaged for execution time of each query utilizing the SFC index and joining it to the uniform 10K data set for each N dimension. Average time of execution for using the Moore SFC index against the uniform 10K data set is shown in *Fig. 6* for the 50 point data queries and *Fig. 9* for the two and four dimensional range queries.

D. Morton Code Space–Filling Curve

The Morton curve or "Z–curve," is a variant from the Peano Curve [1] of 1890. When G. M. Morton decided to analyze geodetic data [4], what we call Geographical Information Services today, he changed the way in which computer science and cartography worked together forever. In his work from 1966 [4], he ushered in a new approach to solving the difficult problem of how to take large areas of unstructured data and bring quantification to its meaning.

We use three algorithms (*MSFC, ToBinary,* and *Inter-Leave*) to compute the SFC index values for a Morton Curve as shown below. These are implemented in Java 8 running on a 2.7 GHz Intel i7 CPU with 16GB RAM. The data is stored and analyzed with PostgreSQL 9.5, and the database connection library used is PostgreSQL JDBC 9.4.1208. The Morton Curve starts with the coordinates of a point in N dimensional space. Each coordinate is converted to a binary number, then interleaved into a new binary number, finally, that binary number is converted into the index value (*vKey*) [16] The process is multi–stepped for implementation of dimensions with data. First, the data is generated (See above.), then imported to tables in the PostgreSQL database. Next, the Java code is executed against the data and a Morton *vKey* is computed for each data point read from the database. The execution time for each dimensional data point (1K, 10K, 100K, 1M) is logged and used to compare against the other N dimension and grouped by quantity of data points (see RESULTS, *Fig. 7*). Our comparisons show each dimension using both uniform and clustered data. The three algorithms below describe the details of these generations. Once the *vKey* is generated and stored into a look–up table, the data is in place for testing.

The Morton Curve testing is done by using a 50 data point query in all N dimensions and four MBR 50 point queries in two and four dimensions joined against the corresponding N dimension uniform 10K data set. Results are averaged for execution time of each query utilizing the SFC index and joining it to the uniform 10K data set for each N dimension. The average execution time for using the Morton SFC index against the uniform 10K data set is shown in *Fig. 7* for the 50 point data queries and *Fig. 10* for the two and four dimensional range queries.

Algorithm 1 Morton Space-filling Curve

1: MSFC()   ▷ Generates index point for 'n' dimension.
2: *iDim* ←I   ▷ $I=\{2,4,8,16\}$
3: *iNumValues* ←D   ▷ $D=\{1K,10K,100K,1000K\}$
4: $dim_{1,2,...,I}$ ←DB
5: *sTime* ←SystemTime
6: **while** $iDim \neq 0$ **do**
7:   $dim_{1,2,...,I}$ ←ToBinary($dim_{1,2,...,I}$)
8: **end while**
9: *vKey* ←InterLeave($dim_{1,2,...,I}$)
10: DB ←vKey, *sTime*, SystemTime, I
11: **return**

Algorithm 2 ToBinary

1: ToBinary($dim_{1,2,...,I}$)
2: *sBin* ←""
3: **while** $dim_{1,2,...,I} > 0$ **do**
4:   *sBin* ← ($dim_{1,2,...,I}$) mod 2
5:   $dim_{1,2,...,I}$ ← ($dim_{1,2,...,I}$) / 2
6: **end while**
7: **return** *sBin*

Algorithm 3 InterLeave

1: InterLeave($dim_{1,2,...,I}$)
2: **for** *<b* in *BitDepth>* **do**   ▷ Bits for each $dim_{1,2,...,I}$
3:   **for** *<i* in $dim_{1,2,...,I}$ó **do**
4:     *vKey* ← (vKey + *dim[b][i]*)
5:   **end for**
6: **end for**
7: **return** *vKey*

III. RESULTS

The experiments were created for measurements of SFC *vKey* index generation speeds for point and range queries as well as utilization of the *vKey* as a look–up and joined to the original point and range data. All experiments were executed in main memory. The validation of these experiments were based on synthetic data sets for two, four eight, and sixteen dimensions. Two categories based on data distribution; uniform and clustered data were performed and the execution time T_i was recorded. Once all the T_i were recorded, their average values were compared for the three SFC as show in *Fig. 11*, *Fig. 12*, and *Fig. 13*.

A. Interpretation of Point Queries Results

1) Hilbert: Fig. 5 The results for the 50 data point query for 10K data points comparing between using Hilbert Curve as an index against Brute-force technique indicates that querying data using Hilbert Curve index is 67% faster than the Brute-Force technique which is a direct result from the reduction in query search space, due to the introduction of the Hilbert Curve indexing. The increase in dimensions has a linear increase in execution time for both of the techniques. These results are observed to be consistent for all N dimensions.

2) Moore: Fig. 6 The results for the comparison between the Moore Curve and brute force on 50 points queries, shows that the Moore Curve performed slightly better than the brute force, except for in sixteen dimensions. This may be the result of calculating the index in a higher dimension. This is a result of the recursion in the algorithm to determine index in multi–dimensional space. The time and CPU memory needed to generate that long of a index value is the cause for the execution time in sixteen dimension.

3) Morton: Fig. 7 The results for the 50 data point query against the uniform 10K data table shows two metrics. The first is brute force and it was performed where each query was individually executed and run time, T_i, was captured, stored into the database, and finally averaged. The second is using the Morton SFC *vKey* as a join to the uniform 10K data table. This reduction in execution time is significant, showing the properties of indexing with database queries.

B. Interpretation of Range Query Results

1) Hilbert Curve: The results for this comparison between two dimension MBRs and four dimension MBRs gave astoundingly increased average execution time for four dimensions, which have approximately 10^3 times more average execution time than two dimensional range queries. This can be attributed due to the increased search space with up to 28M data points that were hit in the *TABLE 2* for 4 dimensional range queries. Fig. 8 is displayed on a logarithmic scale due to this escalation in execution time.

2) Moore Curve: Fig. 9 The comparison results of Moore Curve between two dimension MBRs and four dimension MBRs shows that the MBRs in two dimension performed better than the MBRs in four dimensions. The results is not unexpected since the algorithm must determine all the index values of each point within the MBR range and then join against the Moore SFC uniform 10K data set and join against original 10K uniform data set in order to return back set of points within the MBR that were hit *TABLE 2*. In addition, the size of MBR 0 and 2 is about 2000% more than the largest MBR in two dimension, therefore increasing the computation time.

3) Morton Curve: Fig. 10 The results for this comparison between two dimension MBRs and four dimension MBRs revealed unexpected results. The Morton algorithm with four dimensional MBRs executed faster than the two dimensional MBRs. This can be explained perhaps by the discovered range of points in the selected MBRs and how many intersected with the SFC itself. For example, the MBR–0 in *TABLE 1* indicates that only 7.36% of the numbers generated intersected with the SFC. In *TABLE 2* with the four dimensional data, the percentage of intersection is even smaller. This indicates why the MBR queries in four dimensions had a smaller execution time. The algorithm to determine if one of the generated points is in an MBR_i is as follows: first the query point's SFC *vKey* value is created, then used as part of a multiple column join between the uniform 10K data points, the uniform 10K SFC *vKey* data, and the current MBR point's *vKey*. If a match is not found between the generated SFC *vKey* and the uniform 10K SFC *vKey* table, no further join to the uniform 10K data points table is made. Further analysis would prove helpful in understanding the MBR higher dimensional properties and behavior for searching when mapped with SFC *vKey* values.

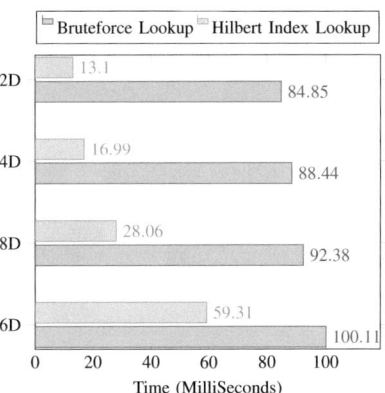

Fig. 5.   Point Query Time - Hilbert

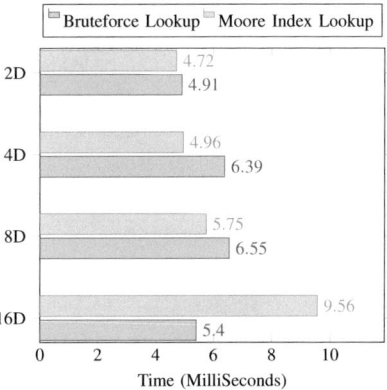

Fig. 6.   Point Query Time - Moore

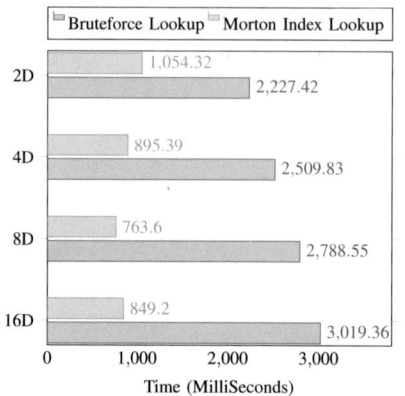

Fig. 7. Point Query Time - Morton

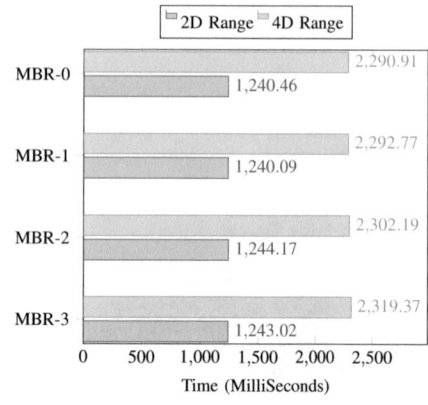

Fig. 9. Range Query Time - Moore

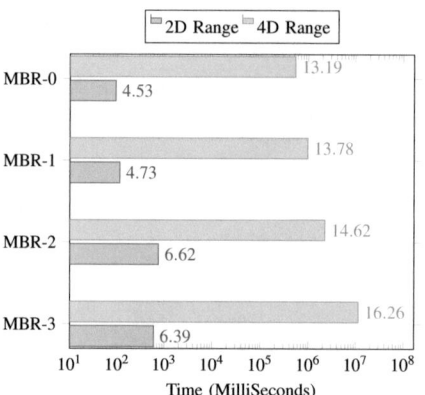

Fig. 8. Range Query Time - Hilbert

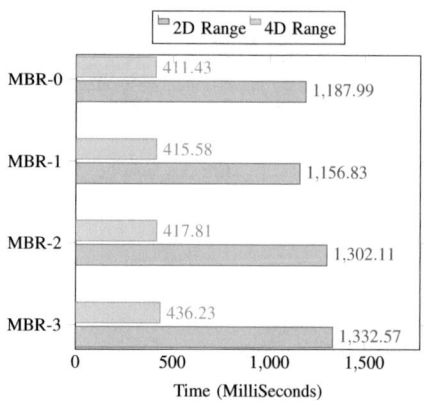

Fig. 10. Range Query Time - Morton

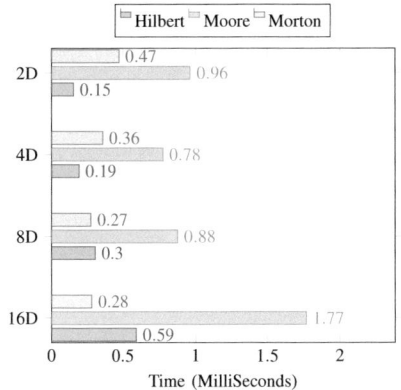

Fig. 11.   Comparative Point Query Time

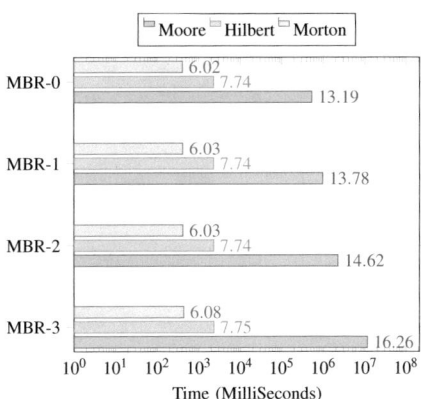

Fig. 13.   Comparative Range Query Time - 4D

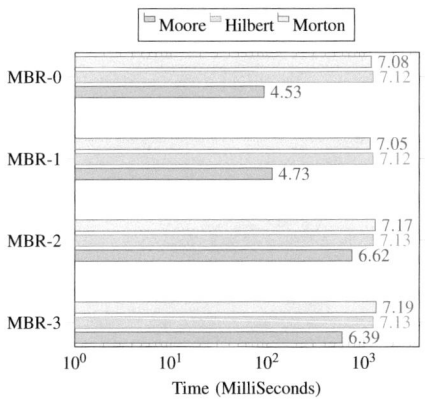

Fig. 12.   Comparative Range Query Time - 2D

IV. FUTURE WORK

Future work includes implementation of numerous kNN queries to demonstrate the effect of execution time relative to the path each SFC creates, produce clustered data for the higher dimensions and larger quantities, implementation of one Sierpinski SFC algorithm in Java to compare to the SFC discussed in this paper, and selectively choose range query MBRs that are similar in size across two, four, eight and sixteen dimensions. Production of metrics in these areas will reinforce our motivation to show high dimension SFC implementations and comparisons.

APPENDIX

TABLE I
TWO DIMENSIONAL RANGE QUERY POINTS

MBR	Generated	Hits
MBR–0	648	88
MBR–1	736	113
MBR–2	5140	786
MBR–3	4092	620

TABLE II
FOUR DIMENSIONAL RANGE QUERY POINTS

MBR	Generated	Hits
MBR–0	11494848	23
MBR–1	4369810	13
MBR–2	10153836	19
MBR–3	2311065	4

TABLE III
HILBERT - UNIFORM SFC GENERATION TIME

Dimensions	1K	10K	100K	1M
2D	195.564	1910.366	21413.774	219174.526
4D	304.412	2512.802	29736.376	312197.349
8D	451.554	4381.562	50556.746	524014.064
16D	993.142	9861.318	119785.806	1211420.218

TABLE IV
HILBERT - CLUSTER SFC GENERATION TIME

Dimensions	1K	10K	100K
2D	199.236	1838.151	30613.498
4D	262.989	2792.002	37084.975
8D	440.067	4434.172	51931.292
16D	1073.014	9862.619	123872.179

TABLE V
MOORE - UNIFORM SFC GENERATION TIME

Dimensions	1K-U	10K-U	100K-U	1M-U
2D	1873	7532	77790	650523
4D	2064	8307	86036	723920
8D	2223	10780	113547	1023488
16D	3581	20739	247613	2338320

TABLE VI
MOORE - CLUSTER SFC GENERATION TIME

Dimensions	1K-C	10K-C	100-C
2D	1880	7372	78585
4D	2063	8322	84652
8D	2468	10754	124479
16D	3990	22025	237142

TABLE VII
MORTON - UNIFORM SFC GENERATION TIME

Dimensions	1K	10K	100K	1M
2D	616	4748	52323	569208
4D	509	4836	57730	597407
8D	553	5234	63207	648693
16D	793	6528	77044	766498

TABLE VIII
MORTON - CLUSTER SFC GENERATION TIME

Dimensions	1K	10K	100K	1M
2D	616	4748	52323	569208
4D	509	4836	57730	597407
8D	553	5234	63207	648693
16D	793	6528	77044	766498

REFERENCES

[1] G. Peano, "Sur une courbe, qui remplit toute une aire plane," *Mathematische Annalen*, vol. 36, no. 1, pp. 157–160. [Online]. Available: http://dx.doi.org/10.1007/BF01199438

[2] D. Hilbert, "Ueber die stetige abbildung einer line auf ein flächenstück," *Mathematische Annalen*, vol. 38, no. 3, pp. 459–460. [Online]. Available: http://dx.doi.org/10.1007/BF01199431

[3] E. H. Moore, "On certain crinkly curves," *Transactions of the American Mathematical Society*, vol. 1, no. 1, pp. 72–90, 1900. [Online]. Available: http://www.jstor.org/stable/1986405

[4] G. M. Morton, *A computer oriented geodetic data base and a new technique in file sequencing.* International Business Machines Company New York, 1966. [Online]. Available: https://domino.research.ibm.com/library/cyberdig.nsf/0/0dabf9473b9c86d48525779800566a39?OpenDocument

[5] T. Ouni, A. Lassoued, and M. Abid, "Gradient-based space filling curves: Application to lossless image compression," in *Computer Applications and Industrial Electronics (ICCAIE), 2011 IEEE International Conference on.* IEEE, 2011, pp. 437–442.

[6] W. Pamuła, "Advantages of using a space filling curve for computing wavelet transforms of road traffic images," in *Image Analysis and Processing, 2003. Proceedings. 12th International Conference on.* IEEE, 2003, pp. 250–253.

[7] T. Zäschke, C. Zimmerli, and M. C. Norrie, "The ph-tree: A space-efficient storage structure and multi-dimensional index," in *Proceedings of the 2014 ACM SIGMOD International Conference on Management of Data*, ser. SIGMOD '14. New York, NY, USA: ACM, 2014, pp. 397–408. [Online]. Available: http://doi.org/10.1145/2588555.2588564

[8] G. Schrack and L. Stocco, "Generation of spatial orders and space-filling curves," *Image Processing, IEEE Transactions on*, vol. 24, no. 6, pp. 1791–1800, 2015.

[9] H. V. Jagadish, B. C. Ooi, K.-L. Tan, C. Yu, and R. Zhang, "idistance: An adaptive b+-tree based indexing method for nearest neighbor search," *ACM Trans. Database Syst.*, vol. 30, no. 2, pp. 364–397, Jun. 2005. [Online]. Available: http://doi.org/10.1145/1071610.1071612

[10] K. G. Pillai, L. Sturlaugson, J. M. Banda, and R. A. Angryk, *Big Data: 29th British National Conference on Databases, BNCOD 2013, Oxford, UK, July 8-10, 2013. Proceedings.* Berlin, Heidelberg: Springer Berlin Heidelberg, 2013, ch. Extending High-Dimensional Indexing Techniques Pyramid and iMinMax(θ): Lessons Learned, pp. 253–267. [Online]. Available: http://dx.doi.org/10.1007/978-3-642-39467-6_23

[11] H. V. Jagadish, "Linear clustering of objects with multiple attributes," *SIGMOD Rec.*, vol. 19, no. 2, pp. 332–342, May 1990. [Online]. Available: http://doi.org/10.1145/93605.98742

[12] M. F. Mokbel, W. G. Aref, and I. Kamel, "Performance of multi-dimensional space-filling curves," in *Proceedings of the 10th ACM International Symposium on Advances in Geographic Information Systems*, ser. GIS '02. New York, NY, USA: ACM, 2002, pp. 149–154. [Online]. Available: http://doi.org/10.1145/585147.585179

[13] A. Butz, "Alternative algorithm for hilbert's space-filling curve," *IEEE Transactions on Computers*, vol. 20, no. 4, pp. 424–426, 1971.

[14] J. K. Lawder, "Calculation of mappings between one and n-dimensional values using the hilbert space-filling curve," *School of Computer Science and Information Systems, Birkbeck College, University of London, London Research Report BBKCS-00-01 August*, 2000. [Online]. Available: http://www.dcs.bbk.ac.uk/TriStarp/pubs/JL1.00.pdf

[15] M. Bader, *Space-Filling Curves: An Introduction With Applications in Scientific Computing*, ser. Texts in Computational Science and Engineering. Springer Berlin Heidelberg, 2012. [Online]. Available: https://books.google.com/books?id=eIe_OdFP0WkC

[16] S. Shekhar and S. Chawla, *Spatial databases: a tour.* Prentice Hall Upper Saddle River, NJ, 2003, vol. 2003.

YOUR KNOWLEDGE HAS VALUE

- We will publish your bachelor's and master's thesis, essays and papers

- Your own eBook and book - sold worldwide in all relevant shops

- Earn money with each sale

Upload your text at www.GRIN.com and publish for free